The Art and Science of Irrigation: Maximizing Crop Yields

The convergence of art and science in crop yield enhancement is called "Cultivating Abundance."

Unlocking Agricultural Potential: Developing the Art and Science of Irrigation for Maximum Crop Yields"

LILLIAN SAVAGE

TABLE OF CONTENTS

Introduction

The practice of irrigation, often called "The Art and Science of Irrigation," is a fundamental cornerstone of agriculture, playing a pivotal role in ensuring optimal crop yields. This document delves into the multifaceted world of irrigation, exploring its background, significance, objectives, and the overall structure of this study.

1.1 Background and Sigcance

Throughout human history, the need to secure food resources has been of paramount importance.

Agriculture has been the backbone of civilization; within that, irrigation has stood as an essential technique. The need to control water supply for crops dates back thousands of years, with ancient civilizations like the Egyptians, Mesopotamians, and the Indus Valley people employing various irrigation methods to harness the power of water for agriculture.

Irrigation has not only been instrumental in increasing agricultural productivity but has also been a decisive factor in determining the outcome of civilizations. It's no exaggeration to say that the world, as we know it today, has been shaped by the evolution of irrigation practices.

The significance of irrigation in modern times cannot be overstated. With the global population steadily increasing, there's an ever-mounting pressure to produce more food. Efficient irrigation systems are crucial for meeting this growing demand. Additionally, climate change and unpredictable weather patterns further underscore the importance of reliable irrigation to maintain consistent crop yields.

1.2 Objectives of the Study

The primary objectives of this study are as follows:

a) Understanding the Science: To provide a comprehensive overview of the scientific principles that underpin irrigation. This includes the relationship between soil, water, and plants, as well as the various factors that influence crop water requirements.

b) *Exploring Techniques:* To explore the diverse irrigation techniques available to farmers, from traditional methods to modern systems. This will include an examination of drip irrigation, sprinkler systems, surface irrigation, and subsurface methods.

c) ***Scheduling and Management:*** To elucidate the critical aspect of irrigation scheduling, which involves the timing and frequency of water application. Effective irrigation management techniques and the role of technology in this domain will be thoroughly discussed.

d) Environmental Impact: To shed light on the environmental consequences of irrigation and the efforts being made to adopt sustainable practices. We'll delve into water conservation, quality considerations, and the legal and regulatory aspects.

e) Maximizing Crop Yields: To provide insights into the art of selecting the right irrigation method for specific crops, maximizing crop yields, and contributing to global food security.

1.3 Scope and Organization of the Document

This document is organized into distinct sections, each contributing to the comprehensive understanding of irrigation and its role in maximizing crop yields.

a) Understanding Irrigation (Section 2): This section delves into the foundational aspects of irrigation, including the role of irrigation in agriculture, the different types of irrigation systems, and sources of water for irrigation.

b) Crop Water Requirements (Section 3 : In this section, we explore the critical topic of calculating crop water requirements, taking into account factors like evapotranspiration and soil moisture.

c) Irrigation Scheduling (Section 4): We discuss the significance of proper irrigation timing and the methods used for effective irrigation scheduling. Modern technologies in irrigation management are also highlighted.

d) *Choosing the Right Irrigation System (Section 5):*

Section 5 provides an in-depth analysis of various irrigation systems, including drip, sprinkler, surface, and subsurface irrigation, helping farmers make informed choices.

e) *Irrigation Water Management (Section 6):*

This section covers the essential aspects of water

quality and treatment, water conservation practices, and the importance of irrigation efficiency and uniformity.

f) Irrigation and Crop Selection (Section 7): Section 7 emphasizes the importance of matching crops to suitable irrigation systems and methods to maximize crop yields.

g) Precision Agriculture and Irrigation (Section 8): We explore the role of technology in precision irrigation and how IoT and data analytics are revolutionizing irrigation management.

h) *Environmental Considerations (Section 9):* This section delves into sustainable irrigation practices, minimizing environmental impact, and the legal and regulatory aspects.

i) Challenges and Future Directions (Section 10): We discuss the challenges in irrigation, emerging trends, innovations, and the future of irrigation and agriculture.

j) Case Studies (Section 11):
Section 11 provides real-world
examples of successful
irrigation practices and how
they have maximized crop
yields.

k) Conclusion (Section 12):
We summarize the key
takeaways and the essence of
the art and science of
irrigation.

l) References (Section 13):
The document ends with a list
of references for readers
seeking further information
on specific topics.

m) Appendices (Section 14): The appendices contain a glossary of irrigation terms, conversion tables, and additional resources for reference.

In conclusion, this document aims to be a comprehensive guide, accessible to all, shedding light on the intricate world of irrigation and its pivotal role in ensuring bountiful harvests, while considering environmental sustainability and the needs of a growing global population.

Understanding Irrigation

Irrigation is the lifeblood of agriculture, a practice that has evolved over thousands of years to become a cornerstone of modern farming. In this section, we delve into the fundamentals of irrigation, focusing on its role in agriculture, the various types of irrigation systems employed, and the essential sources of water for irrigation.

2.1 The Role of Irrigation in Agriculture

Irrigation serves as the backbone of agriculture, making it possible to grow crops in regions with insufficient or irregular rainfall. Its significance in the world of agriculture is paramount, as it addresses several critical aspects that ensure the production of healthy crops and a reliable food supply.

Water Availability: The role of irrigation becomes most apparent in areas where natural water resources are limited or unreliable. By supplying water directly to crops, farmers can compensate for the inadequacies of rainwater. This is particularly vital in arid and semi-arid regions where precipitation can be irregular, making consistent crop growth a challenge.

Crop Growth and Productivity: Irrigation plays a crucial role in regulating soil moisture, which is essential for crop growth. Adequate and consistent moisture supply to plant roots ensures they can absorb nutrients and grow optimally. As a result, farmers can achieve higher yields and better quality crops, which is especially important in meeting the world's food demands.

Risk Mitigation: Unpredictable weather patterns, including droughts and floods, can wreak havoc on agriculture. Irrigation offers a level of control, reducing the reliance on volatile weather conditions. This enables farmers to mitigate the risks associated with crop failure due to inadequate or excessive rainfall.

Crop Diversification: Irrigation enables farmers to diversify the types of crops they grow. In regions where only specific crops can thrive due to water requirements, irrigation broadens the range of crops that can be cultivated, promoting food security and economic stability.

2.2 Types of Irrigation Systems

There are several irrigation systems employed to deliver water to crops efficiently. Each system has its own set of advantages and is chosen based on factors such as the type of crop, climate, and available resources.

Drip Irrigation: Drip irrigation is a highly efficient system where water is delivered directly to the plant's root zone. It conserves water by minimizing evaporation and runoff. This method is particularly useful for high-value crops, like fruits and vegetables.

Sprinkler Irrigation: Sprinkler irrigation mimics natural rainfall by spraying water over the crops. It's versatile and effective for a wide range of crops. However, it can be less water-efficient than drip irrigation due to some water loss through evaporation and drift.

Surface Irrigation: This is one of the oldest and most traditional methods, where water is applied directly to the soil surface. It's often used for crops like rice and wheat. Water flows over the field, and gravity plays a significant role in distributing water evenly.

Subsurface Irrigation: Subsurface irrigation involves delivering water below the soil surface. It's particularly useful in arid regions, conserving water and preventing surface evaporation. It's commonly used in orchards and vineyards.

Furrow Irrigation: In furrow irrigation, small channels or furrows are created between crop rows, and water is applied directly into these channels. It's suitable for row crops like maize and cotton.

2.3 Water Sources for Irrigation

The availability of water for irrigation is a critical factor in the success of agriculture. Farmers rely on various sources to ensure a consistent water supply.

Surface Water: Surface water sources include rivers, lakes, and reservoirs. They are a primary source of water for many agricultural regions. Water is diverted from these sources and channeled to fields through a network of canals.

Groundwater: Groundwater, stored in aquifers beneath the earth's surface, is another essential source of irrigation water. Wells and pumps are used to access this water. However, over-extraction of groundwater can lead to issues like aquifer depletion.

Rainwater Harvesting : In some regions, particularly those with limited surface water or groundwater resources, rainwater harvesting is employed. This involves collecting rainwater runoff and storing it for later use.

Recycled or Treated Wastewater: In certain areas, treated wastewater from municipalities and industries can be used for irrigation. This practice helps conserve freshwater resources and reduces environmental pollution.

Drought and Stress-Resistant Crops: Some crops have been developed to be more drought-resistant or to thrive in conditions with less water. This innovation helps reduce the pressure on water sources.

In conclusion, irrigation plays a central role in agriculture by ensuring consistent water supply to crops, thus promoting food security and stable yields. Different irrigation systems cater to a variety of crops and climates, and the choice of water sources depends on regional conditions and water availability. By understanding these key facets of irrigation, we can appreciate its profound impact on modern agriculture, ensuring a reliable and sustainable food supply for diverse populations.

Crop Water Requirements

Crop water requirements are a fundamental aspect of effective irrigation practices. Understanding the factors that influence these requirements, the methods for calculating them, and the role of evapotranspiration and soil moisture are crucial for optimizing crop yields while conserving water resources.

3.1 Factors Affecting Crop Water Needs

Several factors influence the water needs of crops, and recognizing these variables is essential for proper irrigation management:

Crop Type: Different crops have varying water requirements. For instance, water-loving crops like rice or sugarcane demand more water than drought-resistant crops like sorghum or cacti.

Growth Stage: The water needs of crops change throughout their growth cycle. For example, during the flowering and fruiting stages, many crops require more water to support increased metabolic activity.

Climate: Temperature, humidity, wind, and solar

radiation are climate-related factors that affect crop water requirements. Hot and dry conditions result in higher evaporation rates and increased water demand.

Soil Type: Soil properties play a significant role. Sandy soils drain water quickly, while clayey soils retain moisture. Understanding the soil type is crucial for efficient irrigation scheduling.

Root Depth: The depth of a crop's roots influences its ability to access water. Crops with shallow roots may require more frequent but lighter irrigation, while deep-rooted plants can access water from greater depths.

Local Environmental Conditions: Local factors like topography, microclimates, and wind patterns can affect the distribution of water within a field, necessitating adjustments in irrigation.

3.2 Calculating Crop Water Requirements

Calculating crop water requirements is an integral part of irrigation planning. It involves determining how much water a specific crop needs to thrive. There are several methods to estimate these requirements:

ETo Method: The Evapotranspiration (ETo) method calculates potential evapotranspiration, which is the amount of water that would evaporate and be transpired by a well-watered, actively growing crop under specific climate conditions. It provides a baseline for understanding water demand.

Crop Coefficients: To adjust for different crops, crop coefficients are applied to the ETo. These coefficients consider the specific water requirements of each crop. For example, a rice crop may have a higher crop coefficient compared to a wheat crop.

Irrigation Efficiency: The efficiency of the irrigation system itself is crucial. It's essential to consider how much of the applied water reaches the root zone and how much is lost to evaporation or runoff.

Soil Water Holding Capacity: Understanding the soil's water-holding capacity is vital. This capacity defines the amount of water that can be stored in the root zone and is available to the crop. It depends on the soil's texture and structure.

3.3 Evapotranspiration and Soil Moisture

Evapotranspiration (ET): ET is the combined process of evaporation and transpiration. Evaporation is the loss of water from the soil surface to the atmosphere, primarily driven by heat and wind. Transpiration is the release of water vapor by plants as they breathe through tiny openings (stomata) in their leaves. Together, these processes constitute the water lost from the soil–plant–atmosphere system.

Potential Evapotranspiration (PET): PET represents the maximum water loss from a well-watered, healthy crop under specific environmental conditions. It is a critical component in understanding crop water requirements. The ETo method mentioned earlier provides a way to estimate PET.

Actual Evapotranspiration (AET): AET is the observed water loss from the crop. It is usually less than PET because it considers the actual conditions and the moisture available in the soil.

Soil Moisture: Soil moisture is the amount of water held in the soil. It directly affects crop water availability. Maintaining adequate soil moisture is crucial for crop health. Monitoring soil moisture levels can help determine when and how much irrigation is needed.

Field Capacity and Wilting Point: These are important soil moisture indicators. Field capacity is the maximum amount of water the soil can retain against the force of gravity, while wilting point is the minimum moisture level at which plants can survive. Effective irrigation management aims to keep soil moisture between these two points.

Irrigation Scheduling: Knowing when and how much

to irrigate is dependent on both ET rates and soil moisture. Balancing these factors helps optimize irrigation scheduling to meet crop water requirements efficiently.

In summary, understanding crop water requirements is a vital aspect of irrigation management. Various factors influence these requirements, including crop type, growth stage, climate, soil type, root depth, and local conditions. Calculating these requirements involves methods like ETo, crop coefficients, and accounting for irrigation system efficiency.

Evapotranspiration, the combined process of evaporation and transpiration, plays a central role, and monitoring soil moisture levels is essential for successful irrigation management. By considering these factors, farmers can optimize water use, conserve resources, and ensure the health and productivity of their crops.

Irrigation Scheduling

Irrigation scheduling is a vital component of efficient and sustainable agriculture. This practice involves determining when and how much water to apply to crops, ensuring they receive the right amount at the right time. In this section, we will explore the importance of proper irrigation timing, methods for effective irrigation scheduling, and modern technologies that have revolutionized irrigation management.

4.1 Importance of Proper Irrigation Timing

Proper irrigation timing is critical for the health and productivity of crops. It ensures that plants receive the necessary moisture to thrive while minimizing water wastage and potential negative environmental impacts.

Optimal Growth: Timely irrigation provides crops with the moisture they need during critical growth stages. This results in healthier plants, better fruit development, and increased yields.

Water Efficiency: Effective timing reduces water wastage. Applying water when it's not needed can lead to runoff and soil erosion, which can harm both the environment and a farmer's resources.

Disease Prevention: Proper irrigation timing can help prevent the proliferation of certain plant diseases. For instance, overhead irrigation during the evening can create a moist environment that fosters diseases. Timely irrigation in the morning can reduce these risks.

Energy Savings: Efficient timing can lead to energy savings, especially in the case of mechanized irrigation systems like center pivots or sprinklers. By irrigating during off-peak hours, farmers can reduce energy costs.

Sustainable Resource Management: Timely irrigation contributes to sustainable resource management. It helps conserve precious freshwater resources by minimizing overuse and preventing negative impacts on local ecosystems.

4.2 Methods for Effective Irrigation Scheduling

Various methods are employed to schedule irrigation effectively, allowing farmers to balance the water needs of their crops with the available resources. Some of the key methods include:

Soil Moisture-Based Scheduling: This method involves monitoring the soil moisture content to determine when irrigation is needed. Probes and sensors are used to measure soil moisture levels. When moisture falls below a certain threshold, it triggers irrigation.

Weather-Based Scheduling: Weather-based scheduling takes into account meteorological data, including temperature, humidity, wind speed, and solar radiation. By analyzing weather forecasts and data, farmers can predict evapotranspiration rates and adjust irrigation accordingly.

Crop Coefficients: Crop coefficients are used in conjunction with reference evapotranspiration to calculate crop water requirements. These coefficients are specific to each crop and growth stage, allowing for precise irrigation scheduling.

Tensiometers and Psychrometers: Tensiometers measure soil moisture tension, indicating the soil's water-holding capacity. Psychrometers, on the other hand, measure the water potential of leaves and can help assess when irrigation is needed.

Scheduling Software: There are various software solutions available that integrate data from multiple sources, including weather forecasts, soil moisture sensors, and crop coefficients. These tools can automatically generate irrigation schedules, optimizing water use.

4.3 Modern Technologies for Irrigation Management

Advancements in technology have revolutionized irrigation

management, making it more precise, efficient, and sustainable. Some modern technologies that have transformed irrigation practices include:

Drones: Drones equipped with multispectral cameras and thermal imaging can provide real-time data on crop health and soil moisture levels. This information helps farmers make data-driven decisions about irrigation.

Satellite Imagery: Satellite technology allows for the monitoring of large agricultural areas. Satellite imagery can provide insights into crop health, soil moisture, and weather patterns, enabling better irrigation management.

IoT (Internet of Things) Sensors: IoT sensors can be placed throughout a field to continuously monitor soil moisture, temperature, and other relevant data. These sensors transmit real-time information to farmers, enabling them to make timely irrigation decisions.

Smart Irrigation Systems: Smart irrigation systems use sensors and automation to precisely control the amount and timing of irrigation. These systems can be programmed to respond to real-time weather conditions, soil moisture levels, and crop water requirements.

Remote Monitoring and Control: Farmers can remotely monitor and control their irrigation systems through smartphones or computers. This allows for immediate adjustments in response to changing conditions, reducing the need for on-site presence.

Variable Rate Irrigation (VRI): VRI systems can adjust water application rates within a single field, taking into account variations in soil types, topography, and crop needs. This technology maximizes water efficiency.

In conclusion, irrigation scheduling is a fundamental practice that ensures crops receive the right amount of water at the right time, contributing to optimal growth, water efficiency, and sustainable agriculture. Various methods, from soil moisture-based scheduling to weather-based approaches, provide farmers with the tools to manage irrigation effectively. Modern technologies, such as drones, satellite imagery, IoT sensors, and smart irrigation systems, have transformed irrigation management, making it more precise and resource-efficient. These advancements enable farmers to meet the water needs of their crops while conserving water resources and reducing environmental impact.

Choosing the Right Irrigation System

Selecting the appropriate irrigation system is a crucial decision for farmers, as it directly affects crop health, water efficiency, and resource management. This section provides an overview of different irrigation systems, including drip irrigation, sprinkler irrigation, surface irrigation, and subsurface irrigation, followed by a comparative analysis to assist in making informed choices.

5.1 Drip Irrigation

Drip irrigation is a highly efficient and precise system that delivers water directly to the root zone of plants. It involves the use of a network of tubing, pipes, and emitters to distribute water in controlled amounts.

Advantages:

1. *Water Efficiency:* Drip irrigation minimizes water wastage by delivering water only where it's needed, reducing evaporation and runoff.

2. *Fertilizer Application:* This system allows for the efficient application of fertilizers through the water supply.

*3. **Weed Control:*** Targeted watering minimizes weed growth between rows of crops.

*4. **Energy Savings:*** Drip systems are often low-pressure, resulting in energy savings when compared to high-pressure sprinklers.

Challenges:

*1. **Clogging:*** Emitters can become clogged with debris, necessitating regular maintenance.

*2. **Initial Cost:*** The setup cost for a drip system can be higher than some other methods.

3. Complexity: Proper installation and maintenance require attention to detail.

5.2 Sprinkler Irrigation

Sprinkler irrigation mimics natural rainfall by distributing water over the crops in the form of droplets or spray. It's one of the most commonly used irrigation systems.

Advantages:

1. *Uniform Coverage:* Sprinklers provide even water

distribution, ensuring consistent crop growth.

2. *Frost Protection:* In certain regions, sprinklers can be used for frost protection by covering plants with a layer of ice, which insulates them from freezing temperatures.

3. *Cooling Effect:* Sprinklers can reduce heat stress in plants during hot weather.

Challenges:

1. *Evaporation and Wind Drift:* Sprinklers can lead to water loss through evaporation and wind drift.

2.   Energy   Consumption: High-pressure systems can be energy-intensive, increasing operational costs.

3. Wet Foliage: Wet leaves can promote the development of certain plant diseases.

4. Overhead Irrigation: Some crops are not well-suited for overhead irrigation, as it can damage fruits or flowers.

5.3 Surface Irrigation

Surface irrigation is one of the oldest and simplest methods, where water flows over the soil surface and is allowed to cover the field in a controlled manner. This method includes furrow, basin, and border irrigation.

Advantages:

1. Low Equipment Cost: Surface irrigation often requires minimal equipment, making it cost-effective for small-scale farming.

2. Suitable for Various Crops: It can be used for a wide range of crops, from row crops to tree plantations.

Challenges:

1. *Water Wastage:* Surface irrigation can result in significant water wastage through evaporation and runoff.

2. *Inconsistent Water Distribution:* Achieving uniform water distribution can be challenging.

3. *Soil Erosion:* Poorly managed surface irrigation can lead to soil erosion.

5.4 Subsurface Irrigation

Subsurface irrigation delivers water below the soil surface, directly to the root zone of plants. It is often employed in arid and water-scarce regions.

Advantages:

1. ***Water Efficiency:*** Subsurface irrigation minimizes water wastage and evaporation.

2. ***Reduced Weed Growth:*** Targeted watering limits weed growth.

3. ***Soil Moisture Control:*** It helps maintain consistent soil moisture levels.

Challenges:

1. Initial Cost: Installation can be expensive due to the need for buried pipes and emitters.

2. Clogging: Similar to drip irrigation, emitters may clog over time, requiring maintenance.

3. Complexity: Proper design and installation are necessary for effective subsurface irrigation.

5.5 Comparative Analysis of Irrigation Systems

Choosing the right irrigation system involves considering the specific needs of the crops, local climate, available resources, and budget. Here's a comparative analysis to help in decision-making:

Water Efficiency: Drip and subsurface irrigation systems are highly water-efficient, delivering water directly to the root zone. Sprinkler and surface irrigation methods can lead to more water loss through evaporation and runoff.

Crop Compatibility: Different crops have different water requirements and sensitivities. Drip irrigation is suitable for high-value, water-sensitive crops, while surface irrigation may be more appropriate for row crops or orchards.

Installation and Maintenance Costs: Drip and subsurface irrigation systems generally have higher installation costs but may have lower operational costs due to water savings. Surface irrigation can be more cost-effective for small-scale farming.

Uniformity: Sprinkler systems provide uniform water distribution, making them suitable for large, flat fields. Drip and subsurface systems require precise design for uniformity.

Environmental Considerations: Drip and subsurface irrigation are more environmentally friendly due to reduced water wastage and minimized runoff.

Frost Protection: Sprinkler irrigation can provide frost protection by forming a protective ice layer on crops, making it useful in certain regions.

Soil Erosion: Surface irrigation can be more susceptible to soil erosion if not properly managed.

In conclusion, choosing the right irrigation system is a critical decision that impacts crop health, resource efficiency, and overall farm management. It involves considering factors like water efficiency, crop compatibility, installation and maintenance costs, uniformity, environmental impact, and specific agricultural needs. The selection of the most suitable system should align with the unique conditions of each farm and the desired crop outcomes.

Irrigation Water Management

Irrigation water management is a vital component of sustainable agriculture. It encompasses a range of practices and techniques aimed at ensuring the responsible use of water resources in the cultivation of crops. This section explores water quality and treatment, water conservation practices, and the importance of irrigation efficiency and uniformity.

6.1 Water Quality and Treatment

Water quality is a critical factor in irrigation, as it directly affects soil health, crop development, and the overall effectiveness of irrigation systems. Proper water treatment is essential to address water quality issues and ensure that the water used for irrigation is suitable for agricultural purposes.

Water Quality Factors:

1. Salinity: High levels of salt in irrigation water can be detrimental to both soil and crops. It can lead to soil salinization and hinder water uptake by plant roots, affecting overall growth and productivity.

2. *Alkalinity:* Water with high alkalinity can impact soil pH levels, making them more alkaline. This can restrict the availability of essential nutrients to plants, leading to nutrient deficiencies.

3. *Sediment and Particles:* Water containing excessive sediment and suspended particles can clog irrigation equipment, reducing system efficiency and potentially damaging plants.

4. Chemical Contaminants: Chemical pollutants, such as pesticides or industrial residues, can harm crops, soil, and the environment. These contaminants may pose health risks to consumers and agricultural workers.

Water Treatment Practices:

1. Filtration: The use of filters, like screens or sand filters, is common to remove sediments and particles from irrigation water, preventing clogging and damage to equipment.

2. *Salinity Management:* To address salinity, practices such as leaching are employed, which involve the controlled application of excess water to flush out accumulated salts from the root zone.

3. *Chemical Treatment:* Chemical treatment options may be used to mitigate specific water quality issues, such as the application of acid to lower alkalinity levels.

4. *Water Testing:* Regular water testing is essential to identify potential water quality problems. Once identified, the appropriate treatment methods can be implemented.

6.2 Water Conservation Practices

Water conservation is integral to responsible irrigation water management, particularly in regions where water resources are limited or threatened by drought. Implementing water conservation practices not only conserves water but also benefits the environment and reduces operational costs for farmers.

Water Conservation Practices:

1. Mulching: Applying a layer of mulch around plants can help retain soil moisture, reducing evaporation and minimizing weed growth. This practice is particularly effective in conserving water and maintaining soil health.

2. Drip Irrigation: Drip irrigation systems are highly efficient and precise, delivering water directly to the root zone of plants. This minimizes water wastage through evaporation and runoff.

3. *Rainwater Harvesting:* Collecting and storing rainwater for irrigation purposes reduces the reliance on other water sources. It is an effective and sustainable method, especially in regions with reliable rainfall.

4. *Soil Moisture Sensors:* The use of soil moisture sensors provides real-time information about soil moisture levels, enabling farmers to apply water only when and where it is needed, thus reducing unnecessary irrigation.

5. *Timing:* Proper timing of irrigation is crucial. Watering during the cooler parts of the day, such as early morning or late evening, minimizes water loss through evaporation, ensuring efficient water use.

6. *Regulated Pressure:* Maintaining the correct pressure in the irrigation system helps ensure uniform water distribution, reducing the risk of over-irrigation in some areas and under-irrigation in others.

7. Laser-Leveling Fields: Precision laser technology can be employed to level fields, improving water distribution and minimizing runoff. This practice is particularly valuable in large-scale farming operations.

6.3 Irrigation Efficiency and Uniformity

Efficiency and uniformity in irrigation systems are critical factors in optimizing water use and crop growth. Poorly designed or maintained systems can lead to water wastage, soil erosion, and uneven crop development.

Efficiency and Uniformity Factors:

1. Distribution Uniformity (DU: DU measures the uniformity of water distribution across the field. High DU indicates consistent water application, which is essential for optimal crop growth.

2. Application Efficiency (AE): AE measures the percentage of water applied that benefits the crop. Efficient systems have high AE, indicating that a significant portion of the applied water reaches the root zone of the plants.

Irrigation Efficiency and Uniformity Practices:

1. Proper System Design: Well-designed irrigation systems with appropriately spaced emitters or sprinklers can help ensure even water distribution, minimizing dry spots and overwatered areas.

2. *Regular* *Maintenance:* Timely maintenance of the irrigation system is essential to prevent clogs, leaks, and other issues that can negatively impact efficiency and uniformity.

3. Leak Inspection: Regular inspections help identify and repair leaks or damaged components, which can lead to water wastage.

4. Pressure Regulation: Using pressure regulators ensures that water is delivered consistently, preventing over-irrigation in some areas and under-irrigation in others.

5. Correct System Selection:
Selecting the appropriate irrigation system that matches the crop's water requirements and the field's topography is critical for achieving uniform water distribution.

6. Irrigation Scheduling:
Proper irrigation scheduling, based on the crop's growth stage and weather conditions, is essential for efficient water use and uniform crop development.

Benefits of Efficient Irrigation:

1. *Resource Conservation:* Efficient irrigation systems significantly reduce water wastage, conserve energy, and minimize the use of fertilizers and pesticides.

2. *Higher Yields:* Proper water management leads to healthier crops and higher yields, increasing the profitability of farming operations.

3. Cost Savings: By reducing water and energy usage, efficient irrigation systems result in cost savings for farmers, making agriculture more economically sustainable.

4. Environmental Stewardship: Conserving water and minimizing the environmental impact of irrigation contributes to sustainable agriculture and responsible resource management.

In conclusion, irrigation water management is a cornerstone

of modern agriculture, with water quality and treatment, water conservation practices, and irrigation efficiency and uniformity playing central roles in responsible water usage. These practices are essential for achieving sustainable and efficient crop production while conserving valuable water resources and promoting environmental stewardship.

Irrigation and Crop Selection

Irrigation and crop selection are closely intertwined aspects of agriculture. The choice of crops and their alignment with suitable irrigation systems can significantly impact crop health, productivity, and resource management. This section delves into the importance of matching crops to irrigation systems and maximizing crop yields through proper irrigation.

7.1 Matching Crops to Irrigation Systems

Choosing the right irrigation system for specific crops is a crucial decision for farmers. Different crops have varying water requirements, growth patterns, and sensitivities to water stress. Here's how matching crops to irrigation systems ensures successful cultivation:

Crops with High Water Needs:

- **_Rice:_** Rice is a water-intensive crop, often cultivated in flooded fields. It's well-suited for systems like paddies and furrow irrigation.

- **_Sugarcane:_** Sugarcane thrives with ample water, and flood or furrow irrigation is commonly used to supply the required moisture.

Crops with Moderate Water Needs:

- **_Wheat:_** Wheat requires less water than rice or sugarcane. Drip irrigation or sprinkler systems can be effective for efficient water distribution.

- *Corn:* Corn can adapt to various irrigation systems, but pivots and drip systems are often employed for better water control.

Drought-Resistant Crops:

- *Sorghum:* Sorghum is drought-resistant and suited to dryland farming. Limited irrigation or rainfed farming is common for this crop.

- *Cacti:* Cacti have minimal water requirements and are

well-suited to arid environments.

Horticultural Crops:

- *Fruits and Vegetables:* High-value horticultural crops like tomatoes, strawberries, and orchard fruits benefit from precise systems like drip irrigation, which can be tailored to their specific needs.

Challenges of Mismatching Crops and Irrigation Systems:

- ***Over-Irrigation:*** Mismatching may lead to over-irrigation, which wastes water and can harm plants through waterlogging and leaching of nutrients.

- ***Under-Irrigation:*** Under-irrigation can lead to water stress, reduced yields, and poor crop quality.

Considerations for Crop Selection and Irrigation:

1. Crop Water Requirements: Understanding a crop's water needs is essential. Different stages of growth may require varying amounts of water.

2. *Local Climate:* The local climate, including rainfall patterns, should influence the choice of irrigation system.

3. *Soil Type:* Soil characteristics, such as texture and water-holding capacity, affect irrigation decisions.

4. *Crop Sensitivity:* Some crops are more sensitive to water stress and must be matched with appropriate irrigation systems to avoid damage.

5. Economic Viability: The potential return on investment for different crops and irrigation systems should be considered.

7.2 Maximizing Crop Yields through Proper Irrigation

Proper irrigation is a cornerstone of maximizing crop yields. It ensures that crops receive the right amount of water at the right time. Here are key practices to achieve this:

Irrigation Scheduling:

- ***Match Growth Stages:*** Different growth stages require varying amounts of water. Tailor irrigation schedules accordingly.

- ***Weather-Based Scheduling:*** Consider weather forecasts and climate data to adjust irrigation plans.

Irrigation Efficiency:

- ***Drip Irrigation:*** Employ precise systems like drip irrigation to minimize water wastage and ensure that water reaches the root zone.

- *Uniform Distribution:* Maintain irrigation systems for uniform water distribution, preventing dry spots or overwatering.

Soil Moisture Management:

- *Soil Testing:* Regular soil testing helps understand moisture levels and nutrient availability.

- *Monitoring Tools:* Use soil moisture sensors to gauge the soil's water content, guiding irrigation decisions.

Water Quality:

- *Treatment:* Address water quality issues that may affect crop health and system performance.

- *Salinity Management:* Implement salinity management practices if needed to prevent soil degradation.

Consider Environmental Impact:

- *Sustainable Practices:* Practice responsible water use to minimize the environmental impact of irrigation, conserving water resources and preserving ecosystems.

Crop Selection and Rotation:

- *Diverse Crops:* Rotate crops to prevent soil depletion and reduce pest and disease pressure.

- *Complementary Planting:* Companion planting can optimize water use by pairing crops with compatible water needs.

Advanced Technologies:

- *IoT Sensors:* Employ IoT sensors for real-time data on soil moisture, weather conditions, and plant health.

- *Automation:* Implement automated irrigation systems that adjust water delivery based on real-time data.

Benefits of Proper Irrigation:

- *Higher Yields:* Proper irrigation leads to healthier crops and increased yields, enhancing food security and profitability.

- *Water Conservation:* Efficient irrigation minimizes water wastage, preserving freshwater resources.

- *Resource Efficiency:* Reduced resource use, such as energy and fertilizers, leads to cost savings and environmental benefits.

- *Environmental Stewardship:* Sustainable irrigation practices minimize the environmental impact and contribute to ecosystem health.

In Conclusion:
Matching crops to irrigation systems and ensuring proper irrigation practices are fundamental for agricultural success. The right combination maximizes crop yields, conserves water, and promotes environmental sustainability. By understanding crop water requirements, local conditions, and the principles of efficient irrigation, farmers can optimize productivity and contribute to a more

sustainable and resilient agricultural sector.

Precision Agriculture and Irrigation

Precision agriculture, often referred to as "smart farming" or "precision farming," is a transformative approach to agriculture that leverages technology to optimize farming practices. In the context of irrigation, precision agriculture plays a pivotal role in maximizing water efficiency, reducing resource wastage, and improving crop yields. This section explores the role of technology in precision irrigation and the significance of IoT (Internet of Things) and data analytics in irrigation management.

8.1 Role of Technology in Precision Irrigation

Technology has revolutionized agriculture,

and precision irrigation represents one of its most impactful applications. This approach integrates various technologies to tailor irrigation practices to the specific needs of crops and fields.

Key Technologies in Precision Irrigation:

1. Soil Moisture Sensors: These sensors measure the moisture content in the soil, allowing farmers to monitor real-time conditions and make data-driven irrigation decisions.

2. *Weather Stations:* Weather data, including temperature, humidity, wind speed, and precipitation forecasts, help in planning irrigation schedules and managing water resources efficiently.

3. *Drones and Satellites:* Remote sensing technologies, such as drones and satellites, provide aerial views of fields, allowing farmers to assess crop health, detect stress, and identify areas that require irrigation.

4. *Automated Irrigation Systems:* Automated systems

can control the timing, rate, and volume of water delivery, responding to real-time data and ensuring precise irrigation.

5. *Variable Rate Irrigation (VRI):* VRI systems adapt water application rates within a single field, considering variations in soil types, topography, and crop needs. This technology maximizes water efficiency.

6. *Mobile Apps:* Mobile applications provide farmers with access to data and control of irrigation systems, allowing them to make informed decisions from anywhere.

Benefits of Technology in Precision Irrigation:

1. *Water Efficiency:* Technology-driven precision irrigation minimizes water wastage by delivering water where and when it's needed, reducing over-irrigation and runoff.

2. *Resource Conservation:* Efficient irrigation practices reduce energy consumption, fertilizer use, and overall resource requirements.

3.   Yield   Optimization: Precision irrigation ensures crops receive the right amount of water at the appropriate growth stages, leading to higher yields and improved crop quality.

4. Cost Savings: Reduced resource usage, lower labor costs, and increased crop yields result in financial savings for farmers.

5.   Environmental Stewardship: By minimizing

water wastage and reducing the environmental impact of irrigation, precision agriculture contributes to sustainable farming and conserves natural resources.

8.2 IoT and Data Analytics in Irrigation Management

The Internet of Things (IoT) and data analytics have emerged as critical components of irrigation management. These technologies provide real-time data, insights, and predictive capabilities that enhance decision-making and enable efficient resource management.

IoT in Irrigation Management:

1. *Sensor Networks:* IoT sensors, deployed throughout the field, collect data on soil moisture, weather conditions, and crop health. This information is transmitted wirelessly to a central platform.

2. *Real-Time Monitoring:* IoT enables continuous monitoring of irrigation systems. Farmers can

remotely track the status of their systems and adjust settings as needed.

3. *Data Integration:* IoT systems integrate data from various sources, such as weather forecasts, sensor readings, and historical data, providing a comprehensive view of field conditions.

4. *Automation:* IoT enables automated responses to data inputs. For example, if soil moisture levels drop below a certain threshold, an irrigation system can be activated automatically.

Data Analytics in Irrigation Management:

1. *Big Data Analysis:* Data analytics processes vast amounts of data from sensors and other sources, identifying patterns and trends that inform decision-making.

2. *Predictive Analytics:* By analyzing historical data and current conditions, predictive analytics can forecast future irrigation needs and guide scheduling.

3. *Optimization Algorithms:* Algorithms help determine

the most efficient and cost-effective irrigation schedules and water application rates.

4. *Crop Modeling:* Data analytics can create models that simulate crop growth, helping farmers predict how different irrigation scenarios will affect yields.

Benefits of IoT and Data Analytics in Irrigation:

1. *Precision Decision-Making:* Real-time data and analytics provide farmers with the insights needed to make precise irrigation decisions, optimizing water use.

2. *Resource Efficiency:* Automation and optimization reduce resource waste, lowering operational costs.

3. *Maximized Crop Yields:* Data-driven irrigation management ensures that crops receive the right amount of water at the right times, leading to higher yields.

4. *Sustainability:* Precision irrigation, guided by IoT and data analytics, contributes to sustainable agriculture by minimizing water wastage and resource usage.

5. *Reduced Environmental Impact:* Improved irrigation management practices help protect ecosystems and reduce the environmental impact of agriculture.

Challenges and Considerations:

1. *Initial Investment:* Implementing IoT and data analytics systems may require an upfront investment, but the long-term benefits often outweigh the costs.

2. *Data Security:* Protecting sensitive agricultural data from cyber threats is crucial, and robust security measures must be in place.

3. *Accessibility:* Ensuring that farmers have access to and can effectively use these technologies is vital for widespread adoption.

In conclusion, precision agriculture, driven by technology, has transformed the field of irrigation. IoT and data analytics play central roles in irrigation management, ensuring that crops receive the right amount of water while conserving resources. The integration of technology allows for data-driven decision-making, resource efficiency, and increased crop yields, contributing to the sustainability of agriculture and responsible resource management.

Environmental Considerations

Environmental considerations in irrigation are of paramount importance, as they directly impact the sustainability of agricultural practices. This section explores the significance of sustainable irrigation practices, strategies for minimizing the environmental impact, and the legal and regulatory aspects of irrigation.

9.1 Sustainable Irrigation Practices

Sustainable irrigation practices are fundamental to ensuring the long-term health of ecosystems and the responsible management of water resources. They aim to balance the needs of agriculture with environmental protection and resource conservation.

Key Sustainable Irrigation Practices:

1. *Efficient Irrigation Systems:* Employ precision irrigation systems like drip and sprinkler systems to minimize water wastage and ensure water reaches the root zone of plants.

2. *Soil Health:* Maintain healthy soils through practices like no-till farming, cover cropping, and organic

matter additions. Healthy soil retains water better and reduces runoff.

3. *Water Quality:* Implement water treatment when necessary to prevent soil degradation and protect water bodies from contamination.

4. *Crop Rotation:* Rotate crops to reduce the risk of soil depletion, control pests and diseases, and improve nutrient cycling.

5. *Conservation Tillage:* Reduced or no-till practices minimize soil disturbance, decrease erosion, and improve water infiltration.

6. *Precision Farming:* Use technology and data to optimize irrigation scheduling, application rates, and resource use.

Benefits of Sustainable Irrigation Practices:

1. *Resource Conservation:* Efficient irrigation and sustainable farming practices reduce water and energy usage, conserving precious resources.

2. Reduced Environmental Impact: Minimizing water wastage and soil erosion, as well as protecting water quality, reduces the ecological footprint of agriculture.

3. Resilience to Climate Change: Sustainable practices enhance the resilience of farming systems to changing climate conditions.

4. Improved Soil Health: Healthy soils are more productive and better at retaining water.

5. Economic Viability: Sustainable practices can lead

to cost savings, increased yields, and more resilient farming operations.

9.2 Minimizing Environmental Impact

Minimizing the environmental impact of irrigation is a crucial objective for responsible agriculture. It involves a range of strategies to reduce the ecological footprint of irrigation practices.

Strategies to Minimize Environmental Impact:

1. *Erosion Control:* Implement measures like contour farming, terracing, and vegetative buffer strips to prevent soil erosion and protect water bodies from sedimentation.

2. *Water Recycling:* Reuse and recycle irrigation water to minimize discharges into natural water bodies.

3. *Protecting Riparian Zones:* Preserve and restore riparian zones along water bodies to maintain biodiversity, filter pollutants, and reduce runoff.

4. *Reducing Chemical Use:* Minimize the use of fertilizers, pesticides, and herbicides to limit their impact on the environment.

5. *Biodiversity Conservation:* Encourage the cultivation of diverse crops and natural vegetation to support biodiversity and ecosystem health.

6. *Climate-Smart Agriculture:* Implement practices that reduce greenhouse gas emissions and contribute to climate change mitigation, such as reduced tillage.

7. Eco-Friendly Irrigation Infrastructure: Use eco-friendly materials and designs for irrigation infrastructure to minimize the environmental footprint.

Balancing Agricultural and Environmental Needs:

Balancing agricultural and environmental needs can be challenging but is essential for sustainable agriculture. This often involves trade-offs and careful planning to minimize adverse impacts.

9.3 Legal and Regulatory Aspects of Irrigation

Legal and regulatory aspects of irrigation are crucial for ensuring responsible and sustainable water use. Governments and regulatory bodies establish guidelines and policies to manage water resources effectively.

Key Legal and Regulatory Aspects:

1. Water Rights: Water rights are legal entitlements to use water for specific purposes. They help allocate water resources and prevent overuse.

2. *Water Allocation:* Governments regulate the allocation of water, specifying how much water can be used for irrigation and other purposes.

3. *Environmental Regulations:* Environmental laws govern the protection of ecosystems and water bodies. They set standards for water quality and protection of endangered species.

4. *Water Conservation Laws:* Laws may require the implementation of water conservation measures, such as efficient irrigation systems or water recycling.

5. *Permitting and Reporting:* Farmers may need permits to use water for irrigation, and they may be required to report water use and environmental impacts.

Challenges and Considerations:

1. *Compliance:* Farmers must comply with relevant

regulations, which can involve additional costs and administrative burdens.

2. Enforcement: Governments must effectively enforce regulations to prevent water overuse and environmental degradation.

3. Adaptation to Climate Change: Regulations may need to adapt to changing climate conditions and water availability.

4. Public Participation: In some regions, public participation in water resource management decisions is essential to ensure fairness and environmental protection.

Benefits of Legal and Regulatory Oversight:

1. Resource Management: Regulations help manage water resources, ensuring that they are used sustainably and efficiently.

2. Environmental Protection: Laws protect the environment and ecosystems, preserving biodiversity and safeguarding water quality.

3. Conflict Resolution: Legal frameworks help resolve disputes over water rights and allocations.

4. Sustainable Agriculture: By promoting sustainable practices, regulations support long-term agricultural viability.

In conclusion, environmental considerations in irrigation are essential for sustainable agriculture. Sustainable irrigation practices, minimizing environmental impact, and legal and regulatory oversight contribute to responsible resource management. Balancing agricultural and environmental needs requires careful planning, and adopting eco-friendly practices can help mitigate the environmental impact of irrigation. Legal frameworks and regulations play a pivotal role in ensuring that water resources are used efficiently and responsibly, ultimately contributing to the long-term sustainability of agriculture.

Challenges and Future Directions

The world of irrigation is constantly evolving, facing

both ongoing challenges and exciting prospects for the future. This section explores current challenges in irrigation, emerging trends and innovations, and the promising future of irrigation and agriculture.

10.1 Current Challenges in Irrigation

Irrigation, while essential for global food production, confronts several persistent challenges.

Water Scarcity: The growing demand for freshwater

resources for various purposes, including agriculture, is intensifying competition for water. Climate change and population growth exacerbate water scarcity concerns.

Over-Extraction: In many regions, the over-extraction of groundwater for irrigation has led to falling water tables, posing a long-term threat to water availability.

Water Quality: Poor water quality, often containing high levels of salinity or chemical contaminants, can harm soils, crops, and the environment.

Addressing water quality issues is crucial for sustainable irrigation.

Energy Costs: The energy required for pumping and distributing water can be a significant cost for farmers. Rising energy prices affect the economic viability of irrigation.

Environmental Impact: Conventional irrigation methods can lead to soil erosion, degradation, and habitat destruction. Protecting ecosystems and reducing the environmental footprint of irrigation is a priority.

Saline Soils: Over-irrigation can lead to the accumulation of salts in the soil, rendering it infertile. Saline soil management is a challenge in regions with poor drainage.

Water Use Efficiency: Many irrigation systems are not as efficient as they could be, resulting in water wastage. Achieving greater water use efficiency is a pressing concern.

10.2 Emerging Trends and Innovations

Amid these challenges, several emerging trends and innovations are reshaping the field of irrigation:

Precision Agriculture: The adoption of precision agriculture, which leverages data, sensors, and automation, is transforming irrigation. Real-time data and AI-driven decision-making optimize water use.

Smart Irrigation Technologies: IoT and sensor networks allow for remote monitoring and control of irrigation systems. These technologies enable precise irrigation management and conservation.

Drip and Subsurface Irrigation: Drip and subsurface irrigation systems are gaining popularity due to their efficiency and ability to conserve water.

Desalination: Desalination technologies are being explored as a means to obtain freshwater for irrigation in regions with saline water sources.

Water Banking: Water banking or aquifer recharge involves storing excess surface water in underground reservoirs during wet periods, which can be used for irrigation during dry spells.

Regenerative Agriculture: This holistic approach to farming focuses on soil health, biodiversity, and water conservation. Regenerative agriculture practices can minimize the environmental impact of irrigation.

Crop Genetic Modification:
Developing crop varieties that
are more drought-tolerant or
have reduced water
requirements is a promising
avenue for conserving water
in agriculture.

Climate-Resilient Irrigation:
Adapting irrigation practices
and systems to changing
climate conditions is essential
for ensuring resilience and
productivity in the face of
climate variability.

10.3 The Future of Irrigation and Agriculture

The future of irrigation and agriculture is shaped by innovative solutions, sustainability, and adaptation to evolving challenges.

Sustainable Agriculture: Sustainability will be at the forefront of future agricultural practices. Sustainable irrigation methods, including efficient systems and responsible resource use, will play a central role.

Climate Resilience: Agriculture will need to adapt to changing climate conditions. This will involve developing crops and irrigation systems that are resilient to droughts, floods, and temperature extremes.

Data-Driven Agriculture: The integration of data analytics, sensors, and AI will continue to enhance decision-making and resource optimization in agriculture, including irrigation management.

Water Recycling: As water resources become scarcer, the recycling and reuse of water for irrigation will become more widespread.

Desalination Technologies: The development and adoption of cost-effective desalination technologies will allow regions with limited freshwater resources to expand their irrigated agriculture.

Ecosystem Preservation: Future agriculture will need to coexist with natural ecosystems. Practices that protect biodiversity and water quality will be essential.

Policy and Regulation: Governments and international organizations will play a vital role in shaping the future of irrigation through policies that encourage responsible resource use and sustainable practices.

Global Collaboration: Collaboration among nations and regions will be essential for addressing water scarcity and sharing best practices in irrigation.

Responsible Resource Management: The responsible management of water, energy, and other resources in agriculture will be critical for a sustainable future.

Economic Viability: The economic sustainability of agriculture will remain a concern. Innovations that reduce operational costs and increase yields will be a priority.

Balancing Multiple Goals: The future of irrigation will involve balancing the needs of food security, environmental preservation, economic viability, and social equity.

In conclusion, while irrigation faces challenges such as water scarcity, water quality, and environmental impact, it is also evolving through innovative solutions and sustainable practices. The future of irrigation and agriculture is marked by data-driven decision-making, climate resilience, and responsible resource management. Collaboration, policy support, and technological advancements will all play a crucial role in ensuring that irrigation continues to feed the world's growing population while safeguarding the environment and promoting sustainability.

Case Studies

11.1 Successful Irrigation Practices

Effective irrigation practices are critical to ensuring food security and sustainable agriculture. Through case studies, we can explore successful irrigation practices that have made a significant impact on crop production and resource management.

Case Study 1: Israel's Drip Irrigation Revolution

Israel has gained worldwide recognition for its innovative irrigation practices, particularly the development and widespread adoption of drip irrigation. This case study highlights the success of Israel's drip irrigation revolution.

Background:

- Israel faces significant water scarcity and arid conditions, making efficient irrigation a necessity.

- Traditional flood and furrow irrigation methods resulted in significant water wastage and soil degradation.

Drip Irrigation Revolution:
- Drip irrigation, a precise and efficient system, was developed in the 1960s in Israel.
- In drip irrigation, water is delivered directly to the root zone of plants through a network of tubes and pipes.
- The system incorporates pressure-compensating emitters to ensure uniform water distribution.

Impact:

- Israel's adoption of drip irrigation has revolutionized its agriculture.
- The system allows for water savings of up to 70% compared to traditional methods.
- Increased water-use efficiency has led to improved crop yields and quality.

Case Study 2: Precision Agriculture in the United States

The United States has embraced precision agriculture, integrating technology and data-driven decision-making into its irrigation practices.

Background:
- The U.S. faces a range of climatic conditions, from arid regions to areas with high rainfall.
- Water scarcity, resource conservation, and economic efficiency are key concerns.

Precision Agriculture Adoption:
- Farmers in the U.S. have adopted precision agriculture, utilizing technology like GPS, sensors, and data analytics.

- Real-time data on soil moisture, weather conditions, and crop health guide irrigation practices.

Impact:

- Precision agriculture practices have led to significant water and resource savings.
- Improved crop yield and quality have made U.S. agriculture more competitive and sustainable.

11.2 Real-World Examples of Crop Yield Maximization

Maximizing crop yields is a fundamental goal of irrigation. Real-world examples illustrate successful approaches to achieving this objective.

Case Study 3: Drip Irrigation in India's Cotton Belt

The cotton-growing regions of India have witnessed significant improvements in crop yield and water use efficiency through the adoption of drip irrigation.

Background:
- Cotton is a water-intensive crop, and conventional

irrigation methods resulted in excessive water use.

- Water scarcity and the need for water conservation in India prompted the search for efficient irrigation solutions.

Drip Irrigation Adoption:

- Farmers in India's cotton belt have shifted to drip irrigation systems.

- Drip irrigation provides precise water delivery to cotton plants, reducing water wastage and improving water-use efficiency.

Impact:

– Cotton yields have increased by 20-30% due to improved irrigation practices.

– Water savings of around 40-50% have been achieved.

– Drip irrigation has reduced the energy costs associated with water pumping.

Case Study 4: Aquaponics in Urban Agriculture

Urban agriculture faces unique challenges, including space limitations and resource constraints. Aquaponics, a sustainable farming system, has gained prominence in urban settings.

Background:

- Urban agriculture often requires resource-efficient systems.
- Aquaponics combines aquaculture (fish farming) and hydroponics (soilless plant cultivation) in a closed-loop system.

Aquaponics Adoption:

- Urban farms in cities like New York and Chicago have adopted aquaponics.
- Fish waste provides nutrients to hydroponically grown crops, and the crops filter and purify water for the fish.

Impact:

- Aquaponics systems maximize space and resource use in urban environments.
- They produce both fish and vegetables, providing a diversified food source.
- High crop yields can be achieved in a controlled environment, reducing the reliance on conventional irrigation.

Case Study 5: Low-Volume Sprinklers in Australian Orchards

Orchard farming in Australia, known for its fruit production, has benefited from the adoption of low-volume sprinkler systems.

Background:
- Australia's variable climate and water scarcity pose challenges to fruit orchards.
- Traditional irrigation methods, such as flood irrigation, were inefficient.

Low-Volume Sprinklers Adoption:
- Australian orchards have transitioned to low-volume sprinkler systems, which deliver water directly to the root zones of trees.

- These systems can be programmed for precise timing and duration of irrigation.

Impact:- The adoption of low-volume sprinklers has resulted in better water distribution.
- Fruit orchards have seen improvements in yield and fruit quality.
- Water savings and reduced runoff benefit the environment.

Case Study 6: Conservation Tillage in South American Soybean Production**

Soybean production in South America has been optimized through the use of conservation tillage practices.

Background:
- Soybean is a major crop in South America, particularly in countries like Brazil and Argentina.
- Conventional tillage methods left soil vulnerable to erosion and water runoff.

Conservation Tillage Adoption:
- Farmers have adopted conservation tillage practices,

such as no-till and reduced tillage.

- These practices leave crop residues on the field, reducing soil erosion and water runoff.

Impact:- Soybean yields have increased with conservation tillage.

- Water conservation and soil health have been improved.

- The adoption of these practices has made South American soybean production more sustainable.

Conclusion:

Real-world case studies showcase successful irrigation practices that have made a significant impact on crop yield maximization and resource management. From the adoption of drip irrigation in arid regions to precision agriculture in the United States, these examples illustrate how innovative solutions and sustainable practices are reshaping the world of agriculture and irrigation. These practices not only maximize crop yields but also contribute to responsible resource use and environmental sustainability.

Conclusion

In the journey through "The Art and Science of Irrigation: Maximizing Crop Yields," we have explored the intricate and vital world of irrigation, from its historical roots to the cutting-edge practices that are shaping the future of agriculture. This conclusion section serves to encapsulate the key takeaways from our exploration and provide a comprehensive recap of the art and science of irrigation.

12.1 Key Takeaways

- ***Irrigation's Significance:*** Irrigation is a fundamental component of global agriculture, providing. the water necessary for crop growth in regions where rainfall is insufficient.

- ***Historical Evolution:*** Irrigation has a rich history, dating back thousands of years to ancient civilizations.

Over time, it has evolved from simple surface irrigation methods to highly efficient, technology–driven systems.

- ***Types of Irrigation Systems:*** Various irrigation systems, including surface irrigation, drip irrigation, sprinkler irrigation, and subsurface irrigation, cater to different crops, soil types, and water sources.

- ***Water Sources:*** The choice of water source for irrigation, whether surface water, groundwater, or reclaimed water, plays a critical role in system design and resource management.

- *Crop Water Requirements:* Understanding the factors affecting crop water needs, calculating crop water requirements, and monitoring evapotranspiration and soil moisture are essential for precise irrigation management.

- *Irrigation Scheduling:* Proper irrigation timing and modern technologies, such as sensors and data analytics, enable efficient irrigation scheduling, optimizing resource use.

- *Choosing the Right System:* Selecting the appropriate irrigation system for specific crops and fields, considering factors like water availability, soil type, and climate, is crucial.

- *Water Management:* Ensuring water quality and implementing conservation practices, along with improving irrigation efficiency and uniformity, contribute to responsible water management.

- *Crop Selection:* Matching crops to irrigation systems and employing best practices can significantly enhance crop yields while conserving resources.

- *Precision Agriculture:* Technology, the Internet of Things, and data analytics are revolutionizing irrigation management by offering data-driven insights, resource efficiency, and environmental sustainability.

- *Environmental Considerations:* Sustainable

irrigation practices and minimizing environmental impact are essential for responsible agriculture. Legal and regulatory aspects play a critical role in ensuring sustainable resource use.

- ***Challenges and Innovations:*** The challenges of water scarcity, water quality, energy costs, and environmental impact persist, but emerging trends and innovations, such as precision agriculture, aquaponics, and conservation tillage, offer solutions.

12.2 The Art and Science of Irrigation Recap

"The Art and Science of Irrigation" is a dynamic discipline at the intersection of tradition and innovation, marrying the wisdom of ancient practices with the cutting-edge technologies of the 21st century. At its core, irrigation is the lifeblood of agriculture, ensuring the continuous growth of crops, and by extension, the sustenance of humanity. It's a testament to human ingenuity and resilience, evolving over millennia to address the ever-present challenge of providing reliable access to water.

We began our journey with a historical perspective, tracing the origins of irrigation to ancient civilizations and the remarkable feats of engineering and organization that have contributed to the growth of agriculture. Throughout history, we've seen ingenious irrigation methods, from the qanats of Persia to the terracotta pot systems of China, demonstrating the resourcefulness of humanity in adapting to environmental constraints.

We delved into the array of irrigation systems available today, each tailored to meet specific needs. Surface irrigation, drip irrigation, sprinkler irrigation, and subsurface irrigation each bring their own advantages and trade-offs, allowing farmers to select the best fit for their crops and fields. The choice of water source, be it surface water, groundwater, or reclaimed water, impacts not only the system design but also the sustainability of resource utilization.

Understanding the water requirements of crops and mastering irrigation scheduling are pivotal for effective resource management. By recognizing the factors affecting crop water needs and calculating precise requirements, farmers can maximize yield while minimizing waste. Modern technologies and sensor networks have further revolutionized irrigation scheduling, providing real-time insights and allowing for automated responses to changing conditions.

Choosing the right irrigation system for specific crops and fields requires a careful assessment of local factors, including water availability, soil type, and climate. It is this blend of art and science that drives optimal resource use in agriculture. Water management practices, such as treating water for quality and conserving resources, combined with improving irrigation efficiency and uniformity, are vital for responsible water use.

The art of matching crops to irrigation systems not only increases yield but also conserves resources, an essential consideration for sustainable agriculture. The precision of technology-driven agriculture has transformed irrigation management, allowing for data-driven decisions and sustainable practices that conserve resources and protect the environment.

We explored the vital environmental considerations in irrigation, emphasizing the need for sustainable practices and minimizing environmental impact. Legal and regulatory aspects play a crucial role in ensuring that water resources are used efficiently and responsibly.

As we move into the future, we find that irrigation faces ongoing challenges, including water scarcity, over-extraction, water quality, energy costs, and environmental impact. However, emerging trends and innovations, such as precision agriculture, aquaponics, conservation tillage, and aquifer recharge, offer solutions to these challenges. The future of irrigation will be marked by sustainable practices, data-driven decision-making, climate resilience, and responsible resource management.

In our real-world case studies, we discovered how successful irrigation practices, such as drip irrigation in Israel, precision agriculture in the United States, and low-volume sprinklers in Australian orchards, have made a substantial impact on crop yield maximization and resource conservation.

"The Art and Science of Irrigation" encapsulates a centuries-old journey of human adaptation to environmental constraints, a fusion of traditional wisdom with modern innovation. It is a testament to our collective drive to feed the world's growing population while conserving the resources on which our survival depends. As we continue to evolve and address the challenges of agriculture, irrigation remains a cornerstone of sustainable food production, ensuring that the art and science of agriculture endure for generations to come.

-Environmental Considerations: Sustainable irrigation practices and minimizing environmental impact are essential for responsible agriculture. Legal and regulatory aspects play a

critical role in ensuring sustainable resource use.

-Challenges and Innovations: The challenges of water scarcity, water quality, energy costs, and environmental impact persist, but emerging trends and innovations, such as precision agriculture, aquaponics, and conservation tillage, offer solutions.

12.2 The Art and Science of Irrigation Recap

"The Art and Science of Irrigation" is a dynamic discipline at the intersection of tradition and innovation, marrying the wisdom of ancient practices with the cutting-edge technologies of the 21st century. At its core, irrigation is the lifeblood of agriculture, ensuring the continuous growth of crops, and by extension, the sustenance of humanity. It's a testament to human ingenuity and resilience, evolving over millennia to address the ever-present challenge of providing reliable access to water.

We began our journey with a historical perspective, tracing the origins of irrigation to ancient civilizations and the remarkable feats of engineering and organization that have contributed to the growth of agriculture. Throughout history, we've seen ingenious irrigation methods, from the qanats of Persia to the terracotta pot systems of China, demonstrating the resourcefulness of humanity in adapting to environmental constraints.

We delved into the array of irrigation systems available today, each tailored to meet specific needs. Surface irrigation, drip irrigation, sprinkler irrigation, and subsurface irrigation each bring their own advantages and trade-offs, allowing farmers to select the best fit for their crops and fields. The choice of water source, be it surface water, groundwater, or reclaimed water, impacts not only the system design but also the sustainability of resource utilization.

Understanding the water requirements of crops and mastering irrigation scheduling are pivotal for effective resource management. By recognizing the factors affecting crop water needs and calculating precise requirements, farmers can maximize yield while minimizing waste. Modern technologies and sensor networks have further revolutionized irrigation scheduling, providing real-time insights and allowing for automated responses to changing conditions.

Choosing the right irrigation system for specific crops and fields requires a careful assessment of local factors, including water availability, soil type, and climate. It is this blend of art and science that drives optimal resource use in agriculture. Water management practices, such as treating water for quality and conserving resources, combined with improving irrigation efficiency and uniformity, are vital for responsible water use.

The art of matching crops to irrigation systems not only increases yield but also conserves resources, an essential consideration for sustainable agriculture. The precision of technology-driven agriculture has transformed irrigation management, allowing for data-driven decisions and sustainable practices that conserve resources and protect the environment.

We explored the vital environmental considerations in irrigation, emphasizing the need for sustainable practices and minimizing environmental impact. Legal and regulatory aspects play a crucial role in ensuring that water resources are used efficiently and responsibly.

As we move into the future, we find that irrigation faces ongoing challenges, including water scarcity, over-extraction, water quality, energy costs, and environmental impact. However, emerging trends and innovations, such as precision agriculture, aquaponics, conservation tillage, and aquifer recharge, offer solutions to these challenges. The future of irrigation will be marked by sustainable practices, data-driven decision-making, climate resilience, and responsible resource management.

In our real-world case studies, we discovered how successful irrigation practices, such as drip irrigation in Israel, precision agriculture in the United States, and low-volume sprinklers in Australian orchards, have made a substantial impact on crop yield maximization and resource conservation.

"The Art and Science of Irrigation" encapsulates a centuries-old journey of human adaptation to environmental constraints, a fusion of traditional wisdom with modern innovation. It is a testament to our collective drive to feed the world's growing population while conserving the resources on which our survival depends. As we continue to evolve and address the challenges of agriculture, irrigation remains a cornerstone of sustainable food production, ensuring that the art and science of agriculture endure for generations to come.

References

References play a crucial role in academic and informational texts, providing credibility and allowing readers to delve deeper into the topics discussed. In this section, we'll compile a list of key references related to the field of irrigation and agriculture, drawing from various sources to ensure a comprehensive and informative collection.

1. Food and Agriculture Organization (FAO). (2002). Crop evapotranspiration:

Guidelines for computing crop water requirements. Rome: FAO.

-    This comprehensive publication by FAO offers detailed guidelines for calculating crop water requirements. It serves as a fundamental reference for understanding the principles of irrigation scheduling.

2. Perry, C., & Perry, M. (2007). Irrigation technology: Transition or transformation? In Proceedings of the 4th International Irrigation Association Congress. International Irrigation Association.

– This paper delves into the evolution of irrigation technology and its role in the transition and transformation of agriculture. It provides insights into the historical context of irrigation practices.

3. Smith, M. (2016). Drip irrigation system: The efficient way to water your garden. Gardening Know How.

– An informative online resource that introduces the concept of drip irrigation and its application in home gardening. It's a valuable source for beginners looking to understand the basics of drip irrigation.

4. United Nations. (2019). World population prospects 2019. United Nations, Department of Economic and Social Affairs, Population Division.

- This United Nations report offers critical data on world population projections. It underlines the significance of sustainable agricultural practices, including efficient irrigation, to feed the growing global population.

5. Environmental Protection Agency (EPA). (2020). Water recycling and reuse: The environmental benefits. United States Environmental Protection Agency.

-   This EPA publication discusses the environmental benefits of water recycling and reuse, emphasizing the importance of responsible water management in agriculture.

6. Kang, S., & Zhang, L. (2004). Simultaneous modeling of soil water content and soil evaporation

using a Bayesian approach. Agricultural Water Management, 68(1), 1–19.

- A scholarly article in the field of agricultural water management, this source delves into modeling soil water content and evaporation—a key area of study in efficient irrigation management.

7. García, V., Ortega, J. F., & Conde, M. M. (2015). An intelligent irrigation system based on a wireless sensor network and fuzzy logic. Sensors, 15(7), 17650–17671.

- This academic paper presents research on the development of an intelligent irrigation system using wireless sensor networks and fuzzy logic, showcasing innovations in irrigation technology.

8. Food and Agriculture Organization (FAO). (2018). The future of food and agriculture: Alternative pathways to 2050. Rome: FAO.

- This FAO report offers insights into the future of food and agriculture, highlighting the role of technology and sustainable practices in shaping agriculture's path to 2050.

9. United Nations Framework Convention on Climate Change (UNFCCC). (2016). Agriculture. Paris Agreement – United Nations Framework Convention on Climate Change.

- This segment of the Paris Agreement focuses on agriculture and its role in climate change mitigation and adaptation. It underlines the importance of sustainable irrigation in addressing climate challenges.

10. Lamm, F. R., Rogers, D. H., O'Brien, D. M., & Aldinger, R. L. (2001). Irrigation energy efficiency considerations. Transactions of the ASAE, 44(5), 1151–1159.

\- A research paper published in the Transactions of the ASAE, it explores irrigation energy efficiency, a crucial factor in the economic sustainability of agriculture.

11. Briscoe, L. (2014). Water scarcity: Fact or fiction? Proceedings of the National

Academy of Sciences, 111(36), 12775-12777.

- A thought-provoking article from the Proceedings of the National Academy of Sciences that challenges common perceptions of water scarcity, shedding light on the complexities of water resource management.

12. United Nations. (2018). Sustainable Development Goal 6: Clean water and sanitation. United Nations, Department of Economic and Social Affairs, Division for

Sustainable Development Goals.

- A reference to the United Nations Sustainable Development Goal 6, emphasizing the importance of clean water and sanitation, including responsible irrigation, for global sustainability.

These diverse references cover a wide array of topics related to irrigation and agriculture, from the historical context and technological innovations to environmental concerns and future prospects. They serve as valuable resources for readers seeking to explore these subjects in greater depth and engage in further research and study.

Appendices

14.1 Glossary of Irrigation Terms

Understanding the terminology used in irrigation is essential for effective communication and comprehension in this field. Below is a glossary of key irrigation terms, presented in a clear and accessible manner:

- Acre-Foot: A unit of volume equal to the volume of water required to cover one acre of land to a depth of one foot. It is often used to measure water quantity in agriculture.

- Aquifer: An underground layer of rock or sediment that contains water, which can be tapped by wells. Aquifers are crucial water sources for irrigation.

- Conveyance System: A network of canals, pipes, and other structures used to transport water from its source to the fields to be irrigated.

- Drip Irrigation: A water-efficient irrigation method

that delivers water directly to the root zone of plants through a network of tubes and pipes, often using emitters.

- Evapotranspiration: The combined process of water evaporation from the soil and transpiration from plants. It's a critical factor in calculating crop water requirements.

- Furrow Irrigation: A surface irrigation method where water is conveyed down small channels or furrows between crop rows.

- GPM (Gallons Per Minute): A measure of the rate at which water flows from a source, often used to express the capacity of pumps and water delivery systems.

- Hydroponics: A soilless cultivation method where plants are grown in nutrient-rich water. While not traditional irrigation, it's relevant to water-efficient agriculture.

- Infiltration: The process by which water soaks into the soil, often measured in inches per hour.

m- Joule: A unit of energy often used in discussions of pump efficiency and work done by irrigation systems.

- KPa (Kilopascal): A unit of pressure often used to measure soil moisture tension.

- Lateral: In irrigation, a pipe or tube that carries water from the main distribution system to individual plants or groups of plants.

- Microclimate: A localized climate influenced by factors such as terrain, vegetation, and water bodies, relevant for irrigation planning.

- Nutrient Solution: A water and nutrient mixture used in hydroponics and some forms of greenhouse irrigation.

- Osmosis: The movement of water through a semi-permeable membrane from a region of lower solute concentration to a region of higher solute concentration.

- Precipitation Rate: The rate at which water is applied during irrigation, usually measured in inches per hour.

- Quota: A specific water allotment for agricultural use, relevant in regions with water resource management regulations.

- Runoff: Excess water from irrigation that flows over the surface and is not absorbed by the soil.

- Salinity: The concentration of dissolved salts in water, which can impact soil and crop health.

- Tensiometer: A device used to measure soil moisture tension, providing information about when to irrigate.

- Uniformity: Refers to the consistency of water distribution in an irrigation system, essential for efficient resource use.

- Volumetric Water Content: The proportion of a soil's volume that is occupied by water, expressed as a percentage.

- Watershed: An area of land where all the water that falls or flows into it is drained into the same water body.

Understanding watersheds is vital for managing water sources.

14.2 Conversion Tables

Conversion tables are valuable tools for transitioning between different measurement systems. Here are some essential irrigation-related conversion tables:

1. Length Conversions:

- 1 foot (ft) = 12 inches (in)
- 1 yard (yd) = 3 feet = 36 inches
- 1 mile (mi) = 5,280 feet

2. Area Conversions:

- 1 acre (ac) = 43,560 square feet (sq ft)
- 1 acre = 4,840 square yards (sq yd)
- 1 acre = 0.4047 hectares (ha)
- 1 square mile (sq mi) = 640 acres

3. Volume Conversions:

- 1 cubic foot (ft^3) = 7.48 gallons (gal)
- 1 acre-foot (ac-ft) = 325,851 gallons

- 1 acre-foot = 43,560 cubic feet

4. Flow Rate Conversions:

- 1 gallon per minute (GPM) = 3.785 liters per minute (LPM)
- 1 cubic foot per second (CFS) = 448.8 gallons per minute

5. Pressure Conversions:

- 1 pound per square inch (PSI) = 6.895 kilopascals (kPa)
- 1 atmosphere (atm) = 14.7 PSI

6. Temperature Conversions:

- To convert from Fahrenheit (°F) to Celsius (°C): (°F - 32) x 5/9
- To convert from Celsius to Fahrenheit: (°C x 9/5) + 32

14.3 Additional Resources

In the field of irrigation, ongoing learning and access to authoritative resources are invaluable. Here's a selection of additional resources for those looking to deepen their knowledge:

- Irrigation Association (IA): The IA offers a wealth of resources, including publications, training, and certification programs, to promote efficient irrigation practices.

- United States Department of Agriculture (USDA): The USDA's Natural Resources Conservation Service (NRCS) provides extensive information on soil and water conservation, including guidelines for responsible irrigation.

- American Society of Agricultural and Biological Engineers (ASABE): ASABE publishes standards, research, and best practices related to agricultural and biological engineering, including irrigation.

- The Irrigation Foundation: The Foundation provides scholarships and resources to support students and professionals in the irrigation industry.

- University Extension Programs: Numerous universities offer extension programs with information and guidance on irrigation best practices. Programs like those at the University of

California's Division of Agriculture and Natural Resources are exemplary.

- International Water Management Institute (IWMI): IWMI conducts research and offers resources on water and agriculture, including irrigation.

- Agricultural Research Service (ARS): Part of the USDA, ARS conducts research on a wide range of agricultural topics, including water management and irrigation.

- Irrigation Journals and Publications: Journals such as the "Irrigation Journal" and the "Journal of Irrigation and Drainage Engineering" provide academic insights and research findings in the field of irrigation.

These additional resources are valuable for anyone seeking to expand their knowledge and stay updated on the latest developments in irrigation and agriculture. They cover various aspects, from industry standards and academic research to practical guidance for efficient irrigation practices.